Henintsoa Jean Baptiste RAMAMINIRINA
Vincent Emile RASAMISON
Stéphan Richard RAKOTONANDRASANA

Ecology and ethnobotany ofAloe macroclada Baker

Henintsoa Jean Baptiste RAMAMINIRINA
Vincent Emile RASAMISON
Stéphan Richard RAKOTONANDRASANA

Ecology and ethnobotany ofAloe macroclada Baker

ScienciaScripts

Imprint

Any brand names and product names mentioned in this book are subject to trademark, brand or patent protection and are trademarks or registered trademarks of their respective holders. The use of brand names, product names, common names, trade names, product descriptions etc. even without a particular marking in this work is in no way to be construed to mean that such names may be regarded as unrestricted in respect of trademark and brand protection legislation and could thus be used by anyone.

Cover image: www.ingimage.com

This book is a translation from the original published under ISBN 978-620-3-45983-8.

Publisher:
Sciencia Scripts
is a trademark of
Dodo Books Indian Ocean Ltd. and OmniScriptum S.R.L publishing group

120 High Road, East Finchley, London, N2 9ED, United Kingdom
Str. Armeneasca 28/1, office 1, Chisinau MD-2012, Republic of Moldova, Europe
Printed at: see last page
ISBN: 978-620-7-15861-4

ACKNOWLEDGEMENTS

First and foremost, I would like to thank God, our Lord, Almighty, who has given us life up to now, and the boundless strength and courage to produce and present this dissertation, the fruit of my work during my three years of study at the *Institut d'Enseignement Supérieur d'Antsirabe Vakinankaratra (IES-AV)*; I would also like to express my sincere thanks to all the people who contributed to the production of this work.

My sincere thanks go to :

❖ Docteur *Habilité à Diriger de Recherches* RAJAONARISON Eddie Franck, *Directeur de Recherches, Directeur de ¡'Institut d'Enseignement Supérieur d'Antsirabe Vakinankaratra (IES-AV),* who agreed to let me present this dissertation.

❖ Doctor RATSIMBASON Michel Alain, *Director of Research, Director of the Centre National d'Application de Recherches Pharmaceutiques (CNARP),* for accepting our internship at this institution.

❖ Doctor RASOLOARINIAINA Jean Robertin, Head of the Environment Department and Head of the Environmental Management Course at the Institut d'Enseignement Supérieur d'Antsirabe Vakinankaratra, who enabled me to continue my training at his establishment.

❖ Professor RASAMISON Vincent Emile, *Teacher at ¡ Institut d'Enseignement Supérieur Antsirabe Vakinankaratra (IES-AV), who is* chairing this presentation of the internship report.

❖ Docteur ANTSONANTENAINARIVONY Ononamandimby, *Maître de Conférences, Enseignant- Chercheur à ¡ Institut d'Enseignement Supérieur d'Antsirabe Vakinankaratra (IES-AV)* who kindly agreed to examine my work.

❖ Doctor RAKOTONANDRASANA Stéphan Richard, *Senior Researcher, Head of the Botany and Ethnobotany Department at the Centre National d'Application de Recherches Pharmaceutiques (CNARP),* for agreeing to supervise my work during months of research.

I would also like to thank the teaching and administrative staff of the *Institut d'Enseignement Supérieur d'Antsirabe Vakinankaratra (IES-AV)* and the other members of the *CNARP Ethnobotany Department* who prepared and taught us throughout, particularly Mr RAKOTOARISOA Marrino, Mr RAKOTONDRAFARA, Andriamalala and Mr RAKOTONDRAJAONA Roland for their guidance, encouragement and scientific advice.

I would also like to thank and express my sincere gratitude to the informers and people of Anjeva, Faratsiho and Miandrarivo for their warm welcome.

Finally, I would not forget to thank my family: my parents, my sisters, my brother and all my close friends, who contributed morally, financially and materially during my years of study and during the writing of this book.

Thank you so much!

SUMMARY

The ethnobotanical survey carried out at the three sites revealed traditional uses for *Aloe macroclada* in the treatment of digestive ailments, certain metabolic diseases, wounds, gonorrhoea and asthma. Its uses in cosmetology and as a means of soil protection were also indicated.

The flora associated with the target species is made up of 44 species belonging to 37 genera and 20 families. The most represented families are : Poaceae, Asteraceae and Fabaceae. In a species-poor zone with open vegetation such as Anjeva, the density is relatively high, but the biomass is low. Where the vegetation is richer and more diverse, as observed at Miandrarivo, the density is low and the biomass high.

An adult individual produces an average of 19 to 40 leaves per plant, varying in size from 40.00 to 64.00 cm long and 10.00 to 12.50 cm wide.

Anthropogenic pressures such as bush fires, the expansion of arable land and illegal harvesting represent a crucial threat to *Aloe macroclada.*

Appropriate measures must be implemented to reduce these harmful human activities.

Key words: *Aloe macroclada,* Anjeva, Faratsiho, Miandrarivo

Table of Contents

INTRODUCTION

Plants are essential to human life, providing raw materials for food, construction, firewood and healthcare. In the area of healthcare in particular, plants are the source of many phytomedicines, medicines and alicaments sold on the market.

According to the WHO, more than 80% of the population in developing countries use traditional medicine (WHO; Sofowora, 2010). In Africa, the majority of people (70-80%) consult traditional practitioners for treatment (Mpondo et *al.,* 2012). In Madagascar, due to economic and socio-cultural factors, the vast majority of the population rely on traditional medicine to meet their primary health needs.

Madagascar is known for its rich flora, with 15,000 species, 83% of which are endemic (Goodman and Benstead, 2005). Although there is not yet an exhaustive list of medicinal species in Madagascar (Rakotoarivelo et *al.,* 2015; Rakotonandrasana, 2013), a number of them are used as medicinal plants. The inventory work of Rabesa et *al* (1990) lists 2774 medicinal plants in Madagascar.

These species are distributed across the different vegetation types that exist in Madagascar, including forests, savannahs and mangroves. However, as a result of current demographic growth and climate change, we are witnessing an alarming deterioration in biodiversity, characterised by a reduction in forest cover, the transformation of natural landscapes and the rarefaction or even disappearance of many indigenous species. Medicinal plants are no exception to these threats. *Aloe macroclada* is one of these threatened species and is the subject of this undergraduate thesis.

Aloe macroclada is a species found in the grassy vegetation of the central highlands. Its leaves are widely used in traditional medicine and are sold both in the markets of major cities such as Antananarivo and in suburban communes (Rabemiafara, 2014). This great importance of *Aloe macroclada* in traditional medicine, which translates into strong market demand, represents a real threat to the plant. It is listed in Appendix III of CITES. The work involved in preparing this report is being carried out at three different sites, namely Anjeva, Faratsiho and Miandrarivo. The overall objective is to contribute to the enhancement and improved scientific knowledge of *Aloe macroclada* by carrying out ethnobotanical and ecological investigations. The specific objectives are to

- Expanding the uses of *Aloe macroclada*
- Characterise the species studied and its habitat at the three sites
- Identify the different pressures on the species studied
- Assess leaf biomass at the three selected sites

The assumptions to be justified are as follows:

- The species associated with the target species vary from site to site.
- The pressures on the target species vary from site to site.
- The biomass of the target species varies from site to site.

This document is divided into three parts. In the first part, we describe the study environment, the materials used and the study methods adopted during the course. The second part presents the results and their interpretation. The last part is devoted to discussions and suggestions, followed by a conclusion.

Chapter I. GENERAL

I. GEOGRAPHICAL LOCATION OF STUDY SITES

The study sites belong to two Regions: Anjeva in the Analamanga Region, and Faratsiho and Miandrarivo in the Vakinankaratra Region. These choices were motivated by the nature of the substrate and the proximity of these sites to major towns.

1.1. ANJEVA

I .1.1. Location of survey site

The survey site is located in the Morarano fokontany, Anjeva rural district, at latitude 16°54'14.9"South, longitude 47°39'50.7"East, at an altitude of 1,300m. This commune is part of the central Merina highland region and is located 25 km east of Antananarivo, in the Analamanga region (Figure 1).

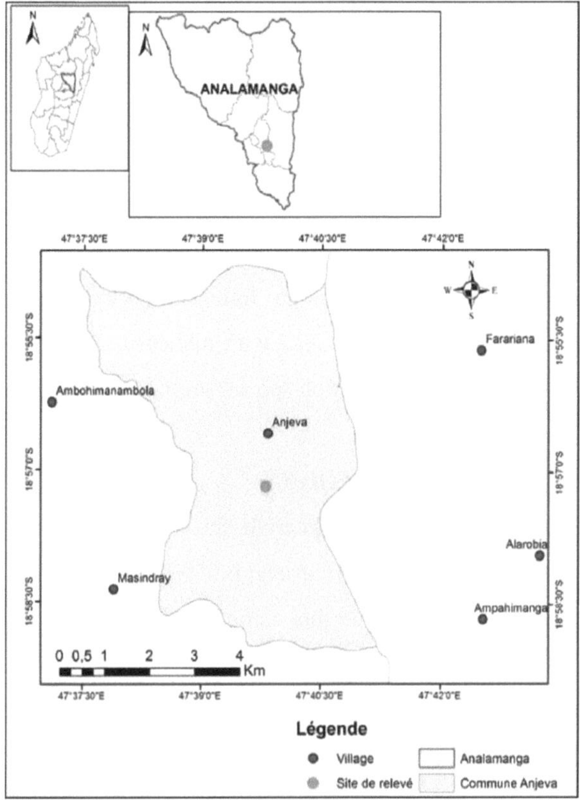

Figure 1: **Location of the Anjeva Antananarivo rural district**

I .1.2. Geology and pedology

Anjeva corresponds to the area occupied by the Carions granite. The hills are made up of more aluminous soils tending towards bauxitisation, while the lower areas are dominated by more fertile grey clays, probably derived from sediments. The soil structure is compact, fragile and difficult to work. They are generally ferralitic (Zebrowski, 2007).

I .1.3. Climate

The climate is tropical, with hot, humid altitudes and two distinct seasons: a hot, rainy season from November to April, and a cool, dry season from May to October. In this commune, rainfall is 1,295 mm, with an annual temperature of 20°C. During the dry season, the average monthly rainfall is 40 mm. Temperatures range from 10°C to 26°C, with an average of 18.5°C in August and 26.7°C in January (ONE, 2007).

I .1.4. Vegetation

On the whole, the vegetation is dominated by grassland formations that have been heavily modified by human activity, cultivated land and reforestation. The grassy formations are found on rocky escarpments and uncultivable areas. They are associated with numerous *Aristida rufescens, Lepturus sp,....* The main reforestation species are *Eucalyptus spp* and *Pinus patula.*

I .1.5. Population and its activities

Part of the province of Antananarivo, the local population belongs to the Merina ethnic group. There are around 6,650 people (ONE, 2007). The main activity of the population in this commune is agriculture. They grow irrigated rice and crops on the hills. Food crops include maize and sweet potatoes, followed by cash crops such as vegetables, carrots and peas.

I.2. FARATSIHO AND MIANDRARIVO

I.2.1. Location of survey site

The surveys were carried out in the Fokontany d'Ambohimandroso, Commune urbaine Faratsiho and in the Fokontany Ambarinomby, Commune urbaine Miandrarivo, District of Faratsiho, Région Vakinankaratra (Figure 2). This District is located 84 km north-west of Antsirabe and Miandrarivo is 32.6 km west of Faratsiho, Vakinankaratra Region. The geographical coordinates of these sites are respectively: latitude 19°23'59" South, longitude 46°57'00" East, latitude 19°26'00" South, longitude 46°45'00" East.

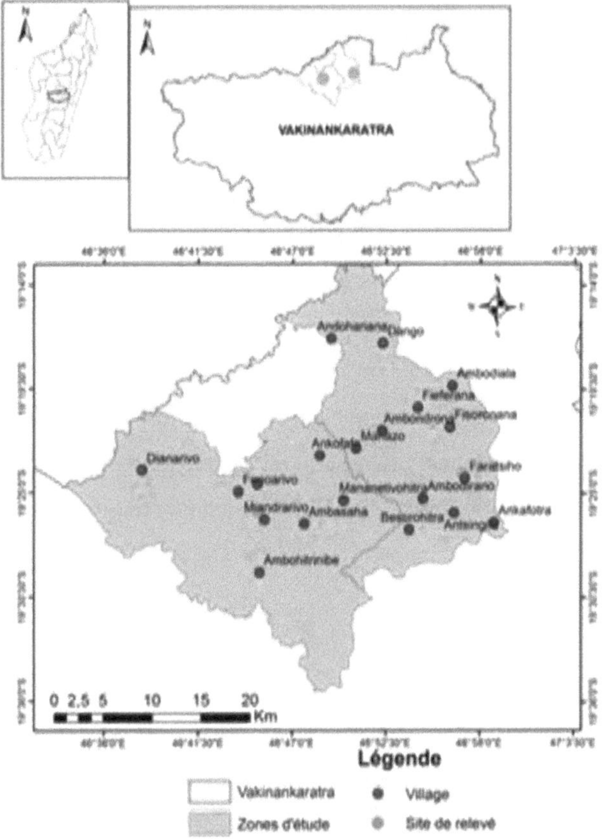

<u>Figure 2:</u> Geographical location of Faratsiho and Miandrarivo

I. 2.2. Geology and pedology

Part of the Ankaratra massif, the geology of the 02 sites is made up of trachytic, ordanchitic, basaltic and ultra-Vulcanic eruptions (Zebrowski, 1971).

As for the soil, there are differences between the two sites:

At Faratsiho, closer to the centre of the massif, the soil is of the volcanic type, more or less deep, of a clayey to clayey-loamy nature, black in colour, and can be friable; as it dries, the elements retract and become very hard (Zebrowski, 1971).

At Miandrarivo, further from the centre, the soil is red ferrallitic, rich in iron and aluminium. The soil structures are compact, fragile and difficult to work. The lower areas are dominated by more fertile grey clays, probably derived from sediments (Zebrowski, 1971).

I. 2.3. Climate

As far as the climate is concerned, we considered the climatic data for the Fartsiho District. The climate is of the high-altitude tropical type. Over the course of the year, the temperature generally ranges from 6°C to 25°C, and rarely falls below 3°C or rises above 28°C. The average annual temperature is 20°C. The average monthly temperature ranges from 7.1°C in August to 26.7°C in January (ONE, 2005). As for rainfall, it rains 122 days a year. Precipitation reaches 1952 mm. There are two seasons: the dry season from May to October and the wet season from October to April (ONE, 2005).

I. 2.4. Vegetation

In Faratsiho District, the vegetation is dominated by grassy and shrubby formations that have been heavily modified by human activities, remnants of Tapia or *Uapaca bojeri* forests, cultivated land and reforestation. The grassy and shrubby formations are found on rocky escarpments and uncultivable areas. They are associated with numerous endemic species, including shrubs such as *Helichrysum gymnocephalum, Helichrysum chermezonii, Vernonia appendiculata, Tetradenia goudotii, etc.,* and herbaceous species such as *Aristida rufescens, Lepturus sp,* etc. Plant formations with *Uapaca bojeri* are degraded sclerophyll forests essentially dominated by *Uapaca bojeri,* associated with *Schefflera bojeri, Rhusta rantana, etc.,* with a discontinuous shrub layer *of Agarista polyphylla, Psiadia altissima* and *Leptolaena pauciflora* (Razafimahatratra, 2018).

Reforestation covers small areas and generally consists of *Eucalyptus spp* and *Pinus patula.*

I. 2.5. Population and its activities

The local population belongs to the Merina ethnic group. The commune of Faratsiho is home to around 165,327 people, while the commune of Miandrarivo has a population of just 27,000 (ONE, 2007). The population farms and raises livestock. The main agricultural activities are irrigated rice-growing and hillside cultivation. Food crops are also important. These include sweet potatoes, potatoes, manioc and maize, as well as cash crops such as vegetables, carrots and peas. There are also fruit trees such as *Prunus persica, Eryobotrya Japonica, Psidium guajava, etc.*

As for livestock farming, the local population raises zebus, pigs, poultry and fish.

II. GENERAL INFORMATION ABOUT ALOE MACROCLADA

II.1 Classification of the species *Aloe macroclada*

Worldwide, the *Aloe* genus comprises more than 624 species (Wendell and Catherine, 1995; Grace et *al.*, 2011), of which around 300 are native to South Africa and more than 151 species are found in Madagascar, where they are all endemic (Carter et *al.*, 2011; Klopper et *al.*, 2013).

According to the APG III (2009) classification system, the plant studied belongs to the :

Phylum : Angiosperms

Class: Monocotyledons Order: Asparagales

Family: Xanthorrhoeaceae Dumort

Subfamily: Asphodeloideae Burnett

Genre: *Aloe*

Species : *macroclada*

Aloe macroclada is known locally as Vahona. It is an endemic species of Madagascar found mainly in the Central Highlands (from Antananarivo to Isalo) and in the south of the island (Anosy region).

II.2 Botanical description

The species was first described by Baker (Perrier de la Bâthie, 1938). It is a succulent plant with leaves 1 to 2 cm thick. Its leaves are sessile and arranged in rosettes of 20 to 48. The ascending, broadly ensiform, arcuate blades, attenuated from the acute base, can reach 75 cm long and 15 cm wide at the middle base. The upper surface is an unblemished soft green, slightly concave at the base and slightly canaliculate towards the top, while the lower surface is convex, similar to the upper surface. The edges are sinuous, toothed to a greater or lesser extent, and fitted with prickly, orange-brown deltoid spines.

The inflorescence reaches a height of 1.75 to 2 m with numerous sterile bracts, the lowest of which are 20 mm long, 35 mm wide and thin, scarce, multi-veined, smaller towards the top (Figure 3).

The flowers are densely grouped around the top of the stem in a large cylindrical spike, 60 to 75 cm long and 6 to 7 cm in diameter. At the base of each flower is an oval bract, acute or cuspid at the top, 10 mm long and 7 mm wide, thin, scarce, with 5 subparallel longitudinal veins. It is borne on a short, thick pedicel 4 mm long and 3 mm in diameter.

The campanulate perianth is formed by 3 outer segments that are free to the base, 5-veined throughout and revolute at the apex at the end of flowering; 3 free inner segments that are wider than the outer ones, keeled, with a more obtuse apex. These segments are pale red inside, greenish outside, 20 to 25 mm long. There are six stamens, each with a lemon-yellow fillet and orange-red anther, 10 mm long. The fruits are capsule-shaped.

Figure 3: Adult plants with inflorescences (a) and *Aloe macroclada* flower (b)

II.3. Uses in traditional medicine

The plant has several virtues:

- ❖ It can be used to treat diseases of the digestive system such as the intestines, liver and stomach, to treat gastric discomfort and as a purgative to eliminate certain intestinal worms (Boiteau, 1979, 1986; Pernet, 1957; Rabearivony, 2010; Rabemiafara, 2014; Razafindrazaka, 2012).

- ❖ With its healing properties, it is used to treat wounds, burns, eczema and skin rashes. It also eliminates dandruff, facial spots, wrinkles, acne and sunburn, and promotes hair growth and smoothing (Desceemacker, 1979; Rabemiafara 2014).

  It is reputed to be effective against certain metabolic diseases such as diabetes, cholesterol, gout, rheumatism, arthritis, arterial hypertension and all forms of tumour and cancer, boosting vigour and providing more energy (Randriamahefa and Rakotozafy, 1979; Rabemiafara, 2014; Razafindrazaka, 2012).

- ❖ The plant is also known to relieve bronchitis, asthma and coughs (Rabemiafara, 2014).

Aloe macroclada is listed in Appendix III of plants traded internationally (CITES, 2002). Because of its fleshy leaves grouped in a rosette, the species is mainly found in public green spaces (Rabemiafara, 2014).

<u>Figure 4 :</u> *Aloe macroclada* in the public garden

Chapter II. MATERIALS AND METHODS

I. DATA COLLECTION AND ANALYSIS

Prior to the fieldwork, orientation sessions on survey methods, floristic data collection and botanical identification were held at the CNARP's Ethnobotany and Botany Department. Bibliographical research on the survey sites and the species studied was also undertaken.

I .1. Ethnobotanical survey

The site visit is carried out during the period from 23 August 2018 until 3 January 2019. Before starting the survey work, a courtesy visit is made to the local authorities. The objectives and duration of our work are explained to them. Then, the search for informants begins. When the informants agree to be interviewed, they are asked about the uses of the target species using the individual survey method. To do this, the plant under study is shown to the informants. The informants cited the uses in traditional medicine, which were recorded on a survey form previously drawn up in accordance with the table (Appendix 1).

Table 1: Survey form

Survey date :

Species name :

Reference herbarium no.

I .2. Inventory of associated flora and density of target species

Before carrying out the floristic survey, a survey of the habitat of the species studied is carried out with a local guide to identify homogeneous vegetation. The aim of this survey is to identify the species associated with the target species. As the species is savannicolous, a 20 m x 8 m plot (Figure 5), divided into 10 4 m x 4 m plots, was used (Braun Blanquet, 1965). All the species present were counted and inventoried. This enables the distribution of species combinations to be observed when the environmental conditions are right (Gounot, 1969).

Similarly, all *Aloe macroclada* individuals present in the plot were counted.

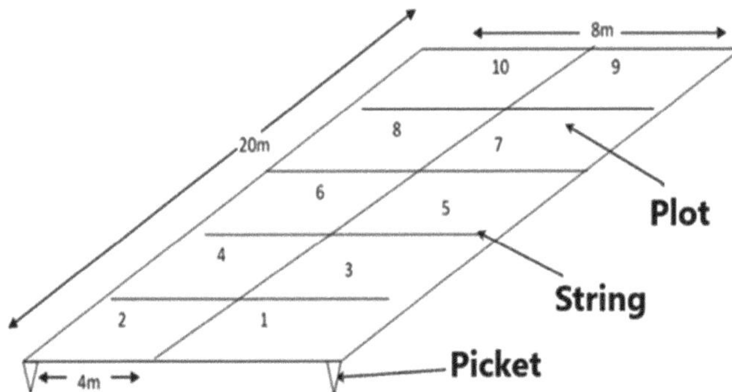

<u>Figure 5:</u> Plot layout using the Braun Blanquet method (1965) for each species surveyed

<u>Figure 6:</u> Flora inventory using the Braun Blanquet method

The frequency of associated species was assessed by passing a decameter stretched at both ends over the herbaceous carpet in the middle of the survey. A sample of 100 points is taken every 20cm (Daget and Poisonnet, 1971), and the frequency is then calculated using the formula :

$$R = b/a * 100$$

Where **R is** the frequency in %, **b** is the number of points where the species is present and **a** is the total number of sample points.

The density of the target species is calculated according to the formula:

$$D = N/S * 100$$

Where **D is** the specific density (ind./ha), **N** is the total number of individuals and **S** is the surface area of the survey plot in ha.

I .3. Creation of reference herbaria

A leafy branch of each plant encountered in the survey plot is taken, then placed in old newspapers that have been numbered beforehand, and finally inserted in the herbarium press. The species' habitat, ephemeral and unobservable characteristics such as colour, odour and habit are noted in a collection notebook. To preserve the herbarium specimens, they are placed in a plastic bag into which a solution of 90° alcohol diluted to 50% is poured until the old newspapers are completely soaked. The plastic bag is taped shut.

These herbarium specimens are dried in the laboratory. To do this, the old newspapers containing the herbarium specimens are inserted in two blotters and two corrugated sheets, put in a herbarium press and introduced into a dryer at 70°C. Daily monitoring is carried out until the specimens are dry.

I .4. Biomass and pressures on the target species

The pressures and threats to *Aloe macroclada* are assessed through direct observations at the three sites and, secondarily, ethnobotanical surveys. These observations were also made at communal markets near the survey site.

To assess biomass, ten adult plants were selected. For each individual, the length and width of the largest leaf were measured. The total number of leaves on each individual was also counted (Figure 7).

Analysis of variance (ANOVA) was used to determine the difference in biomass between the three sites. The parameters measured were the number, length and width of leaves. The concept of the significance of the difference at the 0.05 probability threshold is as follows:

- ❖ The difference is non-significant (NS) when the observed probability of Fisher's test (F) is greater than 0.05.

- ❖ The difference is significant (S) when the observed probability of F is between 0.01 and 0.05.

- ❖ The difference is highly significant (HS) when the observed probability of F is less than 0.01.

XLSTAT 2008 software is used during the analysis

<u>Figure 7:</u> Measuring length and width (a) and counting leaves (b)

Chapter III. RESULTS

I. uses of *aloe macroclada*

There are many uses for *Aloe,* but they vary according to the study sites.

I .1. Cultural uses and soil protection

In Miandrarivo, people believe that *Aloe macroclada* protects the soil from erosion and is used to prevent landslides. This species can stop the formation of "lavaka" (Figure 8).

In all three sites, it is planted in front of or on the door of tombs after a burial to close the tomb. Moreover, for a fairly long period, death no longer attracts another member of the family. Sometimes between dwellings and tombs to separate life and death. To express the succession of joys and sufferings, people compare life with honey and *Aloe* "tantely amam-bahona ny fiainana". It also prevents erosion of the soil deposited above the tomb door.

Figure 8: *Aloe macroclada* **on eroded land (a), at the human habitat (b) and in front of the tomb door (c).**

I .2. Use in traditional medicine

According to the surveys carried out, there are a variety of uses for this species in traditional medicine. *Aloe macroclada* leaves are used to treat at least six (06) illnesses:

1.2.1. Injury and wound

It is used to treat wounds and sores. It is prepared by cutting a leaf in half, then scratching the inside of the leaf and applying the scratching for a day or night to the wounds or lesions. This treatment is continued until the patient is healed.

1.2.2. Diseases of the digestive system

Used internally, the leaves are mixed with honey and a little alcohol to treat digestive tract disorders such as stomachache, pancreas and liver trouble. The plant also has laxative properties.

The resulting preparation should be drunk in three glasses a day. It can also be mixed with sugar or honey.

1.2.3. Gonorrhoea

Take a leaf of *Aloe macroclada,* then scrape the inside to obtain the viscous liquid, then mix with honey and place in a dry place overnight. Drink at least one glass every morning and evening.

1.2.4. Hair care

It promotes hair growth and can accelerate hair growth. To do this, mask your scalp with a gel of this type, leave on for 30 minutes and rinse thoroughly.

1.2.5. Dandruff

It eliminates dandruff. To prepare, mix equal amounts of the gel obtained after scratching with coconut milk and wheat germ oil. Mask the scalp with the mixture and then rinse thoroughly.

1.2.6. Asthma

The leaves are boiled in water for inhalation to relieve asthma. They can cure chronic bronchitis and reduce allergic asthma attacks, without causing any side effects.

II.   SOCIO-ECONOMIC ASPECTS

As far as the socio-economic aspect is concerned, some people live from the sale of *Aloe macroclada* products. They are sold along with other medicinal plants on the town's markets.

II.1 Local sale of *Aloe macroclada*

Aloe macroclada is sold on the market in large cities such as Antananarivo as well as in suburban towns such as Antsirabe. However, they are sold only or together with certain medicinal plants at the roadside in large towns. The number of leaves sold each day is around 17 to 24. The price of a leaf is between 300 and 800 Ariary depending on size, and the liquid product such as gel costs 1200 to 2000 Ariary per half-litre.

Sellers of *Aloe macroclada* leaves on the market get around 1,800 to 3,000 Ariary. In a month, they receive around 90,000 Ariary.

II.2 International sales of *Aloe macroclada*

Internationally, trade is dominated by the cosmetics and pharmaceutical sectors. Products in the form of soap, liquid extracts (shampoo, syrup, etc.), creams and ointments are sold in pharmacies. By Order N°75- 014 of 5 August 1975, Madagascar renewed the CITES convention. The export of ornamental or medicinal plants is regulated by Law No. 2005-018 of 17 October 2005, including *Aloe macroclada*, and is listed in Appendix III.

III. SPECIES ASSOCIATED WITH THE TARGET SPECIES

A total of 44 species in 37 genera and 20 families were recorded in the *Aloe macroclada* habitat at the three sites. The overall list of these species is presented in Appendix 2.

III.   1. Richness of associated flora

The floristic richness of each site is shown in Table 1. This table shows that Miandrarivo is the richest in associated flora, with 42 species in 36 genera and 19 families. Anjeva has the fewest associated species, with 08 species in 08 genera and 05 families inventoried. At Faratsiho, there are 12 associated species in 10 genera and 7 families. No Pteridophyte species were recorded at Anjeva. In each site, Dicotyledons are dominant and are mainly represented by species belonging to the Asteraceae, Fabaceae and Lamiaceae families. Monocotyledons are mainly made up of species from the Poaceae family. The most represented families are the same on all three sites and correspond to the Poaceae, Asteraceae and Fabaceae families.

		Pteridophytes	Monocotyledons	Broadleaf plants	Total
Anjeva	Species	0	3	5	8
	Genres	0	3	5	8
	Families	0	2	3	5
Faratsiho	Species	1	3	8	12
	Genres	1	3	6	10
	Families	1	2	4	7
Miandrarivo	Species	3	7	32	42
	Genres	3	7	26	36
	Families	2	3	14	19

<u>Table 1:</u> Richness of associated flora at the three sites

III.   2. Phytogeographical distribution

The phytogeographical distribution of the species found at the three sites is summarised in Figure 9. Faratsiho is the richest in native endemic species. Among the plants recorded, endemic species predominate (59%), followed by non-endemic native species (25%) and finally introduced species (16%). Miandrarivo is the least rich in endemic species, with only 35%. At Anjeva, endemic species associated with *Aloe macrocada* account for 38%.

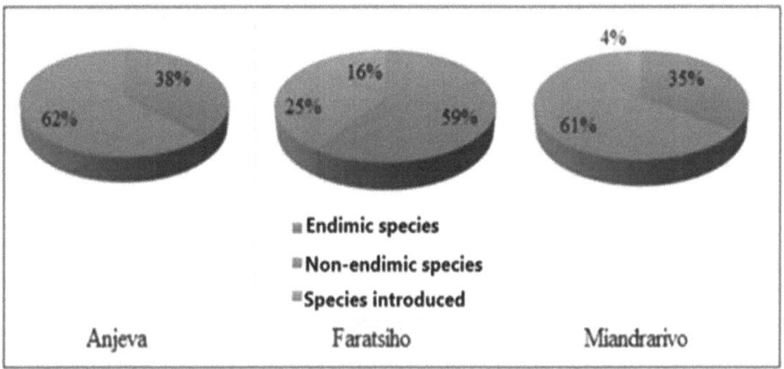

Figure 9: Phytogeographical distribution of species

III.   3. Frequency of associated species

The most frequent species are shown in Table 2. In Anjeva and Faratsiho, the species most frequently associated with *Aloe macrocada* is *Lepturus sp,* with frequency values of 91.42% and 76.08% respectively. This species is not a frequent visitor to Miandrarivo. In this site, *Aristida rufescens* has the highest frequency value at 44.04%. These species are followed by *Lepironia articulata* in Anjeva, *Helichrysum gymnocephalum, Tetradenia goudotii* in Faratsiho and *Kotschya strigosa, Lepironia articulata* in Miandrarivo. Shrubby species are only observed in the habitats of the species in Miandrarivo and Faratsiho. They are represented by *Tetradenia goudotii* and *Helichrysum gymnocephalum* in these two localities and *Billburttia capensoides* in Miandrarivo.

Table 2: Frequency of species associated with *Aloe macroclada*

	Frequency (%)		
Scientific names	Anjeva	Faratsiho	Miandrarivo

Aloe macroclada	4,05	0	2,38
Aristida rufuenscens	0	0	44,04
Bidens pilosa	0	0	3,57

Table 2: Frequency of species associated with *Aloe macroclada*

Elephantopus scaber	0	0	5,95
Helichrysum chermezonii	0	0	2,38
Helichrysum gymnocephalum	0	8,86	3,57
Helichrysum sp	0	2,17	2,38
Hyparrhenia rufa	0	0	11,90
Lepironia articulata	4,28	4,34	9,52
Lepturussp	91,42	76,08	0
Tetradenia goudotii	0	8,86	2,38
Billburttia capensoides	0	0	1,19
Total recovery	74	46	84

(a)   (b)

Figure 10: *Aloe macroclada* **and associated shrubs:** *Tetradenia goudotii* **(a) and** *Helichrysum gymnocephalum* **(b)**

III.4. Density of *Aloe macroclada*

The Anjeva survey site is the richest in terms of numbers of individuals, with 26,250 individuals/ha, while the Faratsiho site has the lowest density, with 8,750 individuals/ha.

Table 3: Density of *Aloe macroclada*

Website	Anjeva	Faratsiho	Miandrarivo
Density (ind./ha)	26 250	8 750	15 000

III.5. Leaf biomass

Aloe macroclada leaf production varies according to the study sites. In general, an adult individual produces 19 to 40 leaves. They measure 85-25cm x 17-8cm.

The analysis of variance of the parameters measured at the three sites was significant. The differences in the number of leaves and leaf length were highly significant (p<0.01) at all three sites. The Miandrarivo plant produced an average of 40 leaves, compared with 21 and 19 leaves for the Anjeva and Faratsiho plants, respectively (Figure 11 a). With regard to leaf length, the average is higher at Miandrarivo (64.00 cm) than at the other sites, with values of 43.00 cm at Anjeva and 40.00 cm at Faratsiho (Figure 11 b).

Variation in plant leaf width (Figure 11 c) was significant (p=0.01). The highest average was found in Miandrarivo (12.50 cm). Leaves are narrower at Anjeva (10.50 cm) and Faratsiho (10.00 cm).

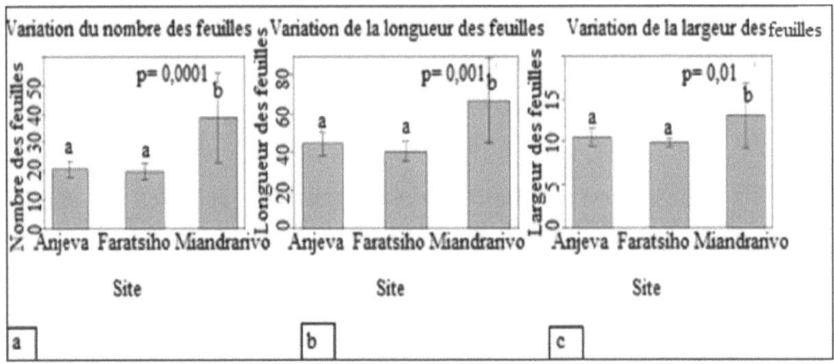

<u>Figure 11:</u> Variation in the number, length and width of leaves

III.6. Threats

The threats to these species are diverse (Table 4) and are all man-made. In general, bush fires, the expansion of agricultural land and deliberate logging are the main causes at the three sites (Figure 12). However, bush fires are very intense at Miandrarivo. Deforestation and slash-and-burn cultivation are dominant at Faratsiho and Miandrarivo. Illegal collection for commercial purposes is threatening the *Aloe macroclada* stands in Anjeva.

(a)   (b)   (c)   (d)

<u>**Figure 12:**</u> **Threats to *VAloe macroclada*: bushfire (a), expansion of agricultural land (b), deliberate felling (c) and sale of medicinal plants (d)** <u>Table 4:</u> **Threats by locality**

Anjeva	Faratsiho	Miandrarivo
Bush fires Extension of agricultural land Voluntary cutting Illegal collection (leaves and seedlings) and marketing	Bush fires Extension of arable land Deforestation and grass cutting Voluntary cutting	Bush fires Extension of arable land Deforestation and cutting of grasses Voluntary cutting

Chapter IV. DISCUSSIONS AND SUGGESTIONS

- **On use:**

Aloe macroclada is a very well-known species, as it is sold both on markets in the capital (Randriamiharisoa, 2015) and on markets in other towns such as Antsirabe (Raherinirina, 2019). Its uses in traditional medicine are very diverse. This diversity of uses is linked to the presence of numerous secondary metabolites such as: aloines, tannins, resins, polysaccharides,.... (Bosch, 2008). The uses in traditional medicine identified during our surveys are confirmed by several authors (Boiteau, 1979, 1986; Desceemacker, 1979; Pernet, 1957; Rabearivony, 2010; Rabemiafara, 2014; Rakotozafy and Randriamahefa, 1979; Razafindrazaka, 2012). In Africa, Aloe species are used primarily to treat malaria (Bjora et *al.*, 2015). This indication is not recorded in our ethnopharmacopoeia despite the high frequency of this disease in Madagascar (Minsan, 2015).

Only the leaves are used. This method of use, combined with local beliefs and use to prevent erosion, could facilitate the sustainable management of this species.

- **Associated flora, leaf biomass and density:**

This study highlights the relationship between leaf biomass and the density of the target species as a function of soil type, associated flora and proximity to the city.

Indeed, leaf biomass is high at sites where the flora is diverse and the associated species include shrub species (Faratsiho and Miandrarivo). This is due to the richness of the soil in organic matter (Razafindrakoto, 2004). Similarly, biomass is low in areas close to the city (Anjeva and Faratsiho). Firstly, the species is sold at the market, where it is highly sought after for its therapeutic properties (Rabemiafara, 2014; Randriamiharisoa, 2015). Traders collect a large number of leaves from an individual in order to earn a sufficient sum of money without taking into account its durability. Thus, collection does not take into account the sustainable management technique, as practised by collectors in other regions of Madagascar (Rakotonandrasana, 2015). Secondly, human population growth requires an increase in arable land to meet food requirements (Erdmann, 2003, Weber, 1994).

The density of individuals at the site with the least diversity of flora and no shrub species (Anjeva) far exceeds those at the two sites. This great disparity could be explained by the high frequency of shrub species at Faratsiho and Miandrarivo, which prevent seeds from germinating. The seeds of this species are light. So, after dissemination, they reach the soil directly and can germinate when conditions are right. On the other hand, at the Miandrarivo site, where the herbaceous layer is very dense, many seeds lose their ability to germinate until they reach the ground. At Faratsiho, the soil is bare as a result of farming and charcoal-making, soil fertility is declining rapidly (Erdmann, 2003), and seeds that fall to the ground are washed away by erosion.

Given the importance of this species and the threats it faces, leaf collectors - both commercial and non-commercial - need to be made aware of the need to preserve it. Indeed, using the plant for generations to come requires planting and sustainable collection practices in order to maintain biodiversity and satisfy human needs. Research into seed germination, the production of higher yielding leaves and harvesting methods is therefore essential.

CONCLUSION

The present work concerns the ecology and ethnobotany *of Aloe macroclada* in the three different sites.

It is a savannah species from the central highlands of Madagascar that is widely used in traditional medicine. The uses identified during this study are confirmed by previous publications. In the three sites selected, the plant is used to treat diseases of the digestive, genitourinary and respiratory systems. It is also used in cosmetology.

The species associated with the target species vary according to the proximity of the town and the nature of the soil. This flora is poor in Anjeva and Faratsiho compared with Miandrarivo. In the first two localities, it is herbaceous species belonging to the Poaceae family: *Lepturus* sp. *Aristida rufescens* in Miandrarivo. Shrub species are rare in Anjeva, and consist mainly of *Helichrysum gymnocephalum* in Faratsiho and *Kotschya strigosa* in Miandrarivo.

Stand density is higher in areas with poor flora. However, production and leaf size in flora-rich areas exceed those in flora-poor areas. This seems to be related to the threats to the species. In fact, in Miandrarivo, a locality far from the town, the individuals have a greater size and number of leaves than in Faratsiho and Anjeva, localities close to the town. What's more, the better-stocked vegetation maintains soil fertility.

Given the various types of anthropic pressure and the importance of *Aloe macroclada* not only culturally, but also socio-economically and environmentally, our outlook is to seek better conditions for seed germination and better ecological conditions for maximum biomass.

BIBLIOGRAPHICAL REFERENCES

1- Angiosperm Phylogeny Group III. 2009. An update of the Angiosperm Phylogeny Group classification for the orders and families of flowering plants: APG III. *Botanical Journal of the Linnaean Society,* 161: 105-121.

2- Bjora C.S., Burki S., Demissew S., Forest F., Grace O.M., Klopper R.R., Ronsted N., Simmonds M.S., Smith G.F., Symonds M.R., Van Wyk A.E., 2015. Evolutionary history and leaf succulence as explanations for medicinal use in aloes and the global popularity of *Aloe verra* BMC Evolutionary Biology, 15-29.

3- Boiteau P., 1979. Dictionnaire des noms malgaches de végétaux, in FITOTERAPIA2, Vol L, 73-96 p.

4- Boiteau P., 1986. Médecine traditionnelle et pharmacopée: précis de matière médicale malgache, 141p.

5- Bosch C.H., 2008. PROTA Network Office Europe, Wageningen University Box P.O., 341, 6700 AH Wageningen, Netherlands. *Aloe arborescens, Aloe buettneri, Aloe divaricata, Aloe ferox, Aloe flexilifolia, Aloe lomatophylloides, Aloe nuttii.*

6- Braun-Blanquet J., 1965. Plant sociology. The study of plant communities. Hafnerpublishingcompany- New York and London.

7- Daget and Poissonet, 1971. Principles of a technique for the quantitative analysis of herbaceous vegetation. Doc. CEPE-CNRS, Vol. 56, 85-100 p.

8- Descheemacker A., 1979, Ravimaintso, 112 p.

9- Erdmann T.K., 2003. The dilemma of reducing shifting cultivation. In The Natural History of Madagascar. Ed. by Goodman S.M., Benstead J.P., 134-139. University of Chicago Press. Chicago and London.

10- Goodman, Benstead, 2006. Updated estimates of biotic diversity and endemism for Madagascar. Oryx, 39: 73-77.

11- Gounot, M., 1969. Méthode d'étude quantitative de la végétation. Masson. Paris.

12- Ministry of Public Health (MINSAN). 2015. Annuaire des statistiques du secteur santé de Madagascar. Service des statistiques sanitaires et démographiques. 99 p.

13- Mpondo E., Dibong D.S, Priso R.J, Ngoye A., Ladoh Y.C.F., 2012. Current status of traditional medicine in the health system of rural and urban populations of Douala Cameroon. Journal of Applied Bio science 55 : 4036-4045

14- World Health Organization. 2002. WHO strategy for traditional medicine 2002-2005, WHO Geneva, 5 p.

15- Pernet R.B., 1957, Les plantes médicinales malgache, Mémoire de l'Institut Scientifique de Madagascar, B., VII, 1-144 p.

16- Perrier de la Bâthie H., 1938. Liliacées Flore de Madagascar, 77-112 p.

17- Rabearivony D.A.N., 2010. Ethnobotanical study of medicinal species in Ambalabe-Vatomandry and evaluation of their ecological status. DEA dissertation. Facultés des Sciences. Université d'Antanarivo, 90 p.

18- Rabemiafara M.V., 2014. Ethnobotany, ecology and conservation status *of Aloe macroclada* Baker in the Analamanga region. DEA dissertation, Ecologie Végétale, Faculty of Sciences, University of Antananarivo, 61 p.

19- Rabesa Z.A., Rabenoro C., Andriantsiferana R., Rakotobe E.A., 1990. *Notes on Malagasy plants used in the traditional pharmacopoeia.* Strasbourg, First International congress on Ethnopharmacology, 5-9.

20- Raherinirina S.N., 2019. Plants sold in the Asabotsy market to treat stomach aches, coughs and wounds. Licence dissertation. Institut d'Enseignement Supérieur d'Antsirabe Vakinankaratra.

21- Rakotoarivelo N.H., Rakotoarivony F., Ramarosandratana A.V.,

Jeannoda V., Kuhlman A.R., Randrianasolo A., Bussmann R.W., 2015. Medicinal plants used to treat the most frequent diseases encountered in Ambalabe rural community, Eastern Madagascar. *Journal of Ethnobiology and Ethnomedicine.* DOI 10.1186/s13002-015- 0050-2.

22- Rakotonandrasana S.R., 2013. Medicinal plants of the Zahamena protected area (Madagascar) and its surroundings: floristic richness and endemicity. *In: Proceedings of the XIXth AETFAT Congress for African Plant Diversity, Systematics and Sustainable Development.* Antananarivo. Edited by N Beau, S Dessein and E Robbrecht. Sciptabotanica Belgica 50:356-362.

23- Rakotonandrasana S.R., 2015. Importance, impacts of the use and rational management of satrana or *Hyphaene coriacea* Gaertn. (Arecaceae). Madagascar conservation en développement, 10: 48-52 p.

24- Randriamahefa M., Rakotozafy A., 1979, Tari-dalàna ahafantarana ny raokandro malagasy, Bokyvoalohany, 406 p.

25- Randriamiharisoa M.N., Kuhlman A.R., Jeannoda V., Rabarison H., Rakotoarivelo N., Randrianarivony T., Rakotoarivony F., Randrianasolo A., Bussmann R.W., 2015. Medicinal plants sold in the markets of Antananarivo, Madagascar. *Journal of Ethnobiology andEthnomedicine.* N, DOI 10.1186/s 13002-015-0046-y.

26- Razafimahatratra T.F., 2018. *Uapaca bojeri* plant formations in the rural Commune of Ramainandro, District of Faratsiho: flora and medicinal plants. Licence dissertation. Institut d'Enseignement Supérieur d'Antsirabe- Vakinankaratra.

27- Razafindrakoto R.M., 2004. Evaluation de l'efficience de diverses techniques biologiques de gestion conservatoire de la fertilité des sols. PhD thesis, University of Antananarivo, 274 p.

28- Razafindrazaka R.M., 2012. Les plantes sauvages les plus utilisées dans la Région Analamanga : Inventaire ethnobotanique dans les communes rurales d'Ankadinandriana, d'Ambohitrandriamanitra et de Miadanandriana. CAPEN dissertation, Ecole Normale Supérieure, Université d'Antananarivo, 70 p.

29- Sofowora A., 2010. Medicinal plants and traditional medicines of Africa. Karthala, 375 p.

30- Tableau de bord environnemental, 2004. Analamanga Region. Office Nationale de l'Environnement, Antananarivo.

31- Tableau de bord environnemental, 2007. Vakinankaratra Region. Office National de l'Environnement, Antananarivo.

32- Weber, J., 1994. *L'occupation humaine des aires protégées à Madagascar : diagnostic et éléments pour une gestion viable* ; CIRAD / ONE - ANGAP - DEF ; Mahajanga. 08p.

33- Zebrowski C., 1971. Properties of Ankaratra andosols. Cah. O.R.S.T.O.M., sér. Pédol, vol. IX, no 1. 83-108p.

APPENDICES

Appendix 1: Survey form

N°	Name	Age	Gender	Profession	Disease treated	Part used	Method of preparation	Associated plant
1								
2								
3								
4								

Appendix 2: Global list of associated species

Gender and species	Family	Common name	Phytogeographical distribution	Presence
Acacia dealbata Link	Mimosaceae	Moza	Introduced	Faratsiho Miandrarivo
Aeschynomene sensitiva Sw.	Fabaceae	Sorindrana	Not endemic	Miandrarivo
Ageratina riparia (Regel) R.M. King et H. Rob.	Asteraceae	Maitsoririnina	Not endemic	Miandrarivo
Aristida rufenscens Steud	Poaceae	Orombohitra	Endemic	Anjeva, Faratsiho Miandrarivo

Bidens pilosa L.	Asteraceae	Tsipolotra	Not endemic	Miandrarivo
Billburttia capensoides Sales et Hedge	Apiaceae	Tsingeraera	Endemic	Miandrarivo

Appendix 2: Global list of associated species

Gender and species	Family	Common name	Phytogeographical distribution	Presence
Blumea crispata (Vahl) Merxm.	Asteraceae	Ariandro	Not endemic	Miandrarivo
Cenchrus polystachios (L.) Morrone	Poaceae	Tsibolo	Not endemic	Miandrarivo
Centella asiatica (L.) Urb.	Apiaceae	Talapetraka	Not endemic	Miandrarivo
Cheilanthes viridis (Forssk.) Sw.	Pteridaceae	Apanga	Not endemic	Miandrarivo
Cynodon dactylon (L.) Pers.	Poaceae	Fandrotrarana	Not endemic	Miandrarivo
Dicranopteris linearis *(Burm. f.) Underw.*	Gleicheniaceae	Apangavy	Not endemic	Faratsiho Miandrarivo
Elephantopus scaber L.	Asteraceae	Parakin' alika	Not endemic	Miandrarivo
Eucaliptus spp Juss.	Myrtaceae	Kininina	Introduced	Miandrarivo
Gerbera elliptica Humbert	Asteraceae	Fotsiavadika	Endemic	Faratsiho Miandrarivo

| *Gladiolus sp* L. | Iridaceae | Glaieul | Not endemic | Miandrarivo |
| *Gomphocarpus fruticosus* (L.) W.T. Aiton | Apocynaceae | Fonorona | Not endemic | Miandrarivo |

Appendix 2: Global list of associated species

Gender and species	Family	Common name	Phytogeographical distribution	Presence
Harungana madagascariensis Lam ex Poir.	Hypericaceae	Harongana	Not endemic	Miandrarivo
Helichrysum chermezonii Humbert	Asteraceae	Rambiazina	Endemic	Miandrarivo
Helichrysum faradifani Scott Elliot	Asteraceae	Ahibalala	Endemic	Faratsiho Miandrarivo
Helichrysum gymnocephalum (D.C.) Humbert	Asteraceae	Rambiazina	Endemic	Faratsiho Miandrarivo
Helichrysum sp Mill.	Asteraceae	Tsatsamaitra	Endemic	Faratsiho Miandrarivo
Hyparrhenia (Nees) Stapf	Poaceae	Orombohitra	Not endemic	Miandrarivo
Indigofera arrecta Hochst. ex A. Rich.	Papilionaceae	Aikavavy	Not endemic	Anjeva Miandrarivo
Indigofera glabra L.	Papilionaceae	Aikalahy	Not endemic	Miandrarivo

Kalanchoe prolifera (Bowie ex Hook.) Raym.-Hamet	Crassulaceae	Sodifafana	Endemic	Miandrarivo
Kotschya strigosa (Benth.) Dewitet P.A. Duvign.	Fabaceae	Tsikobona	Not endemic	Anjeva Miandrarivo

Appendix 2: Global list of associated species

Gender and species	Family	Common name	Phytogeographical distribution	Presence
Lepturus sp R. Br.	Poaceae	Kotofohy	Not endemic	Anjeva Faratsiho Miandrarivo
Kotschya strigosa (Benth.) Dewitet P.A. Duvign.	Fabaceae m.v. Fabaceae h.v	Tsikobona	Not endemic	Anjeva
Lantana camara L.	Verbenaceae	Voanalakely	Not endemic	Miandrarivo
Lepironia articulata (Retz.) Domin	Boroginaceae	Vero	Not endemic	Anjeva Faratsiho Miandrarivo
Physalis peruviana L.	Solanaceae	Voanatsindrana	Not endemic	Miandrarivo
Pinus patula Ten	Pinaceae	Kesika	Introduced	Faratsiho

Psiadia angustifolia (Humbert) Humbert	Asteraceae	Dingadingana	Endemic	Anjeva Miandrarivo
Psiadia altissima (DC.) Drake	Asteraceae	Dingadingana	Endemic	Faratsiho Miandrarivo
Psidium guajava L.	Myrtaceae	Goavy	Not endemic	Miandrarivo
Senecio multidenticulatus Humbert	Asteraceae	Sinisio	Endemic	Anjeva Miandrarivo

Appendix 2: Global list of associated species

Gender and species	Family	Common name	Phytogeographical distribution	Presence
Taraxacum officinale F.H. Wigg.	Asteraceae	Hanitrin'ny patsaka	Not endemic	Miandrarivo
Tetradenia goudotii Briq.	Lamiaceae	Borona	Endemic	Faratsiho Miandrarivo
Vaccinium secundiflorum Hook.	Ericaceae	Varamotsana	Endemic	Miandrarivo
Vernonia appendiculata Less.	Asteraceae	Ambiaty	Endemic	Anjeva Miandrarivo
Vernonia pectoralis Baker	Asteraceae	Sakatavilotra	Endemic	Miandrarivo

<u>**Appendix 3:**</u> Species associated with *VAloe macroclada*: *Psiadia altissima* (a), *Aristida rufenscens* (b), *Mussaenda arcuata* (c), *Kalanchoe prolifera* (d), *Kotschya strigosa* (e) and *Lepturus sp* (f)

<u>**Appendix 4:**</u> Some ways of preparing *Aloe macroclada* in traditional medicine Leaf cut (a), inner surface scraped (b) and leaf mixed with honey and a little alcohol (c)

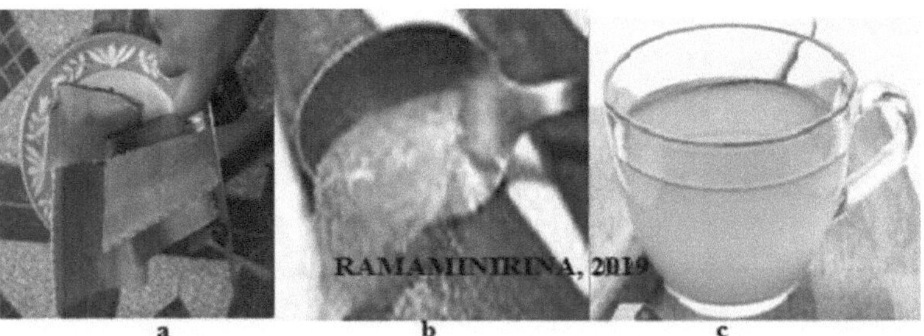

Appendix 5: Some of the threats to *Aloe macroclada*: bush cultivation (a), bush fires (b), deliberate felling (c) and cutting of grasses (d)

a
b
RAMILIMIRINA 2019
c
d

Surname and first name(s): RAMAMINIRINA Henintsoa Jean Baptiste

Dissertation title: **ECOLOGY AND ETHNOBOTANY** *OF ALOE MACROCLADA*: **ANJEVA, FARATSIHO AND MIANDRARIVO**

Pagination : 27

Tables: 04

Figures: 12

Summary

The ethnobotanical survey carried out at the three sites revealed traditional uses for *Aloe macroclada* in the treatment of digestive ailments, certain metabolic diseases, wounds, gonorrhoea and asthma. It has also been used in cosmetology and as a soil protection agent. The flora associated with the target species is made up of 44 species belonging to 37 genera and 20 families. The most represented families are: Poaceae, Asteraceae and Fabaceae. In a species-poor area with open vegetation such as Anjeva, the density is relatively high, but the biomass is low. Where the vegetation is richer and more diverse, as observed at Miandrarivo, the density is low and the biomass high. An adult individual produces an average of 19 to 40 leaves per plant, varying in size from 40.00 to 64.00 cm long and 10.00 to 12.50 cm wide. Anthropogenic pressures such as bush fires, the expansion of arable land and illegal collection represent a crucial threat to *Aloe macroclada.* Appropriate measures must be implemented to reduce these harmful human activities.

Keywords: *Aloe macroclada,* Anjeva, Faratsiho, Miandrarivo

Rapporteur: Doctor RAKOTONANDRASANA Stéphan Richard, Senior Researcher

Author's address: 0116 F 440 Ambohimiandrisoa Antsirabe

E-mail: ramaminirinah7@gmail.com

Contact: 034 61 400 73/ 033 85 078 49

Printed by Books on Demand GmbH, Norderstedt / Germany